Index

1. What is Electricity..3

2. Introduction of AC and DC..4

3. Induction Motor..5

4. Types of an Induction Motor 6

5. Working Principal of a Induction Motor..7

6. Main Components..8

7. Advantages of a Induction Motor..11

8. Induction Motor Troubleshooting..13

9. Basic Induction Motor Troubleshooting Chart....................................17

10. Tips for maintain a Induction Motor...20

11. Precautions for maintenance a Induction Motor..............................18

Start of Book

Electricity:

All matter is made up of atoms, and atoms are made up of smaller particles called protons, neutrons and electrons. The center or nucleus of an atom is made up of protons and neutrons. The electrons orbit around the nucleus just like the moon orbits around the earth. Electrons have a negative charge, while protons have a positive charge and neutrons are neutral (i.e., they have no charge).

Electricity is created when particles become charged. Some are negatively charged (electrons), and some are positively charged (protons). These opposite charges attract, whereas particles with similar charges repel each other.

An electron is two thousand times smaller in mass than a proton, but its electrical charge is equal to that of a proton. Electrons of many elements, particularly metals, are easily knocked off from their parent atoms and can wander freely between atoms. If a state of unbalanced charges exists, then a necessary condition to create an electric current also exists. However, the flow of electric current cannot take place until the circuit is completed.

When an electrical source (such as a battery) is attached by a wire to a form of resistance (such as a light or motor), and a circuit is completed back to the source or to ground, free electrons are released into the wire, creating an electrical potential or voltage. The electrons bounce against other electrons in the wire, which are repelled because they have the same electrical charge. They go on bouncing against other free electrons down the wire, causing a flow of electrons-an electrical current. Provided there is somewhere for the electrons to go and be converted into another form of energy (such as inside a light or motor), the electrons will flow out the far end.

Type of Current:

AC(Alternating Current):

AC is short for **alternating current**. This means that the direction of current flowing in a circuit is constantly being reversed back and forth. This is done with any type of AC current/voltage source.

The electrical current in your house is alternating current. This comes from power plants that are operated by the electric company. Those big wires you see stretching across the countryside are carrying AC current from the power plants to the loads, which are in our homes and businesses. The direction of current is switching back and forth 60 times each second.

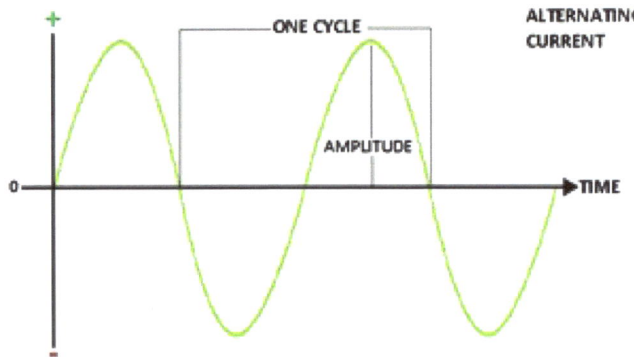

DC(Direct Current)

Direct current (DC) is electrical current which flows consistently in one direction. The current that flows in a flashlight or another appliance running on batteries is direct current.

One advantage of alternating current is that it is relatively cheap to change the voltage of the current. Furthermore, the inevitable loss of energy that occurs when current is carried over long distances is far smaller with alternating current than with direct current.

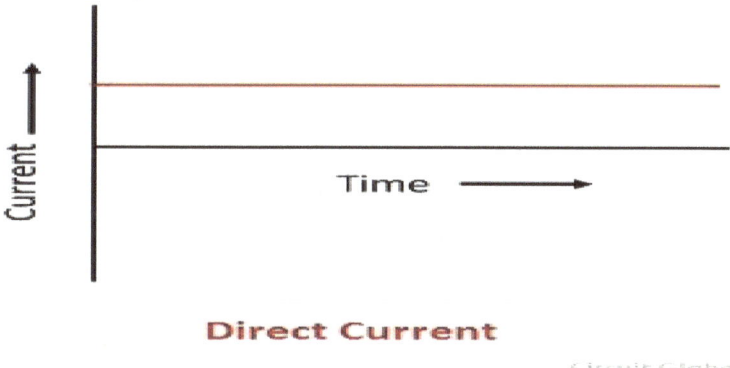

Induction Motor:

Induction Motors are the most commonly used motors in many applications. These are also called as **Asynchronous Motors**, because an **induction motor** always runs at a speed lower than synchronous speed. Synchronous speed means the speed of the rotating magnetic field in the stator.

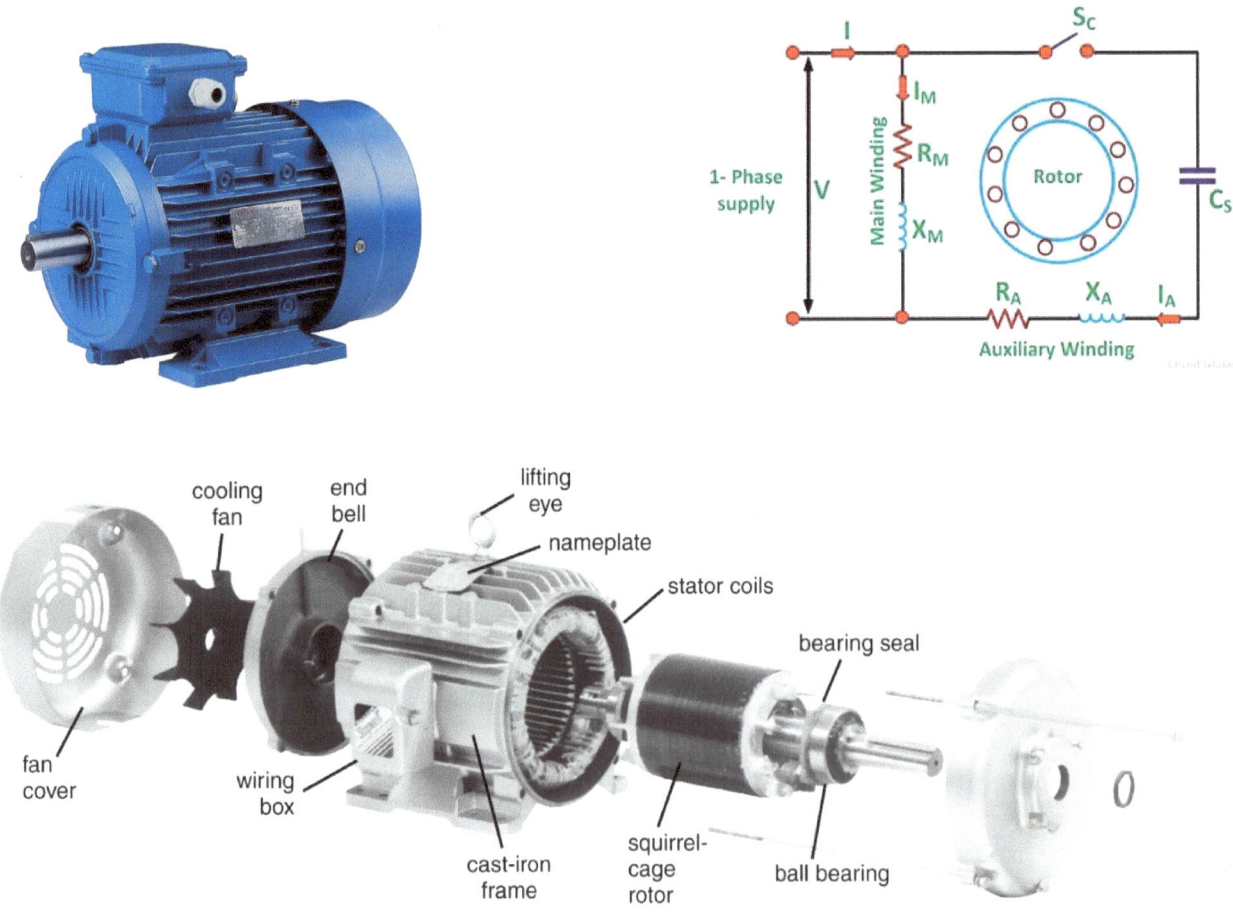

Types of Induction Motor:

There basically 2 **types of induction motor** depending upon the type of input supply –

(i)   Single phase induction motor

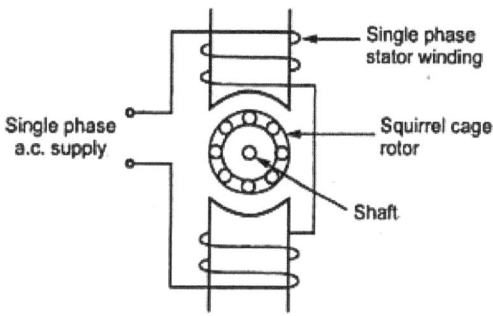

(ii)   Three phase induction motor.

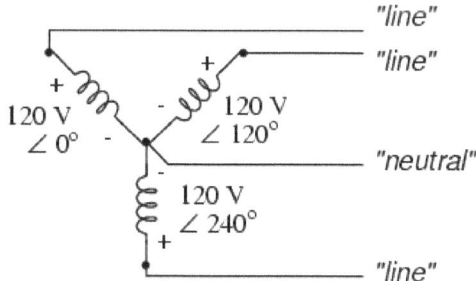

Basic Working Principle of An Induction Motor:

In a DC motor, supply is needed to be given for the stator winding as well as the rotor winding. But in an **induction motor** only the stator winding is fed with an AC supply.

- Alternating flux is produced around the stator winding due to AC supply. This alternating flux revolves with synchronous speed. The revolving flux is called as "Rotating Magnetic Field" (RMF).
- The relative speed between stator RMF and rotor conductors causes an induced Electro Magnetic Field (EMF) in the rotor conductors, according to the Faraday's law of electromagnetic induction. The rotor conductors are short circuited, and hence rotor current is produced due to induced EMF. That is why such motors are called as **induction motors**.
  (This action is same as that occurs in transformers, hence induction motors can be called as **rotating transformers**.)

- Now, induced current in rotor will also produce alternating flux around it. This rotor flux lags behind the stator flux. The direction of induced rotor current, according to Lenz's law, is such that it will tend to oppose the cause of its production.
- As the cause of production of rotor current is the relative velocity between rotating stator flux and the rotor, the rotor will try to catch up with the stator RMF. Thus the rotor rotates in the same direction as that of stator flux to minimize the relative velocity. However, the rotor never succeeds in catching up the synchronous speed. This is the **basic working principle of induction motor** of either type, single phase of 3 phase.

Main Components:

The A.C. Induction Motor has three main parts, rotor, stator and enclosure. The stator and rotor do the work and the enclosure protects the rotor and stator.

STATOR CORE:

The stator is the stationary part of the motor's electromagnetic circuit and is made up of thin metal sheets, called laminations. Laminations are used to reduce energy losses that would result if a solid core was used. Stator laminations are stacked together forming a hollow cylinder to reduce eddy current and hysteresis losses.

Stator Core

STATOR WINDINGS:

Coils of insulated wire are inserted into slots of the stator core. When the assembled motor is in operation, the stator windings are connected directly to the power source. Each grouping of coils together with the steel core it surrounds becomes an electromagnet when current is applied. Induction is the basic principal behind motor operation.

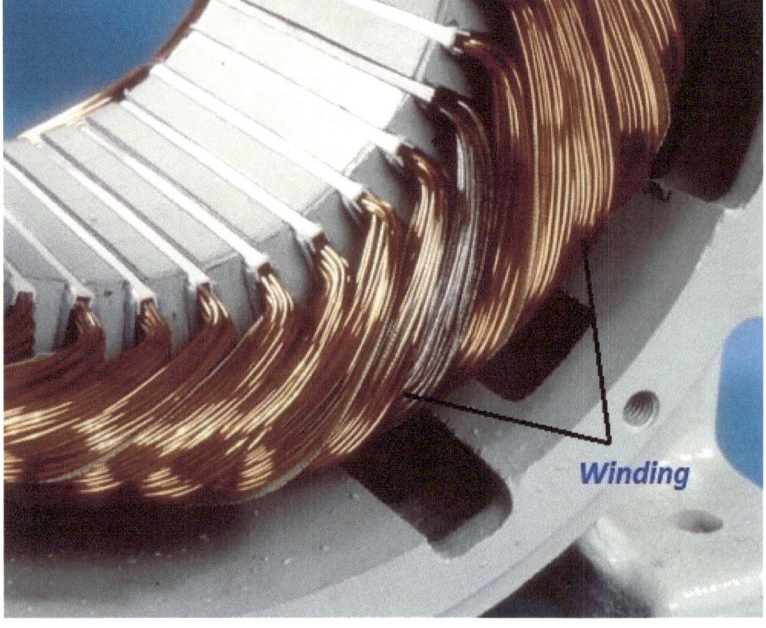

ROTOR CONSTRUCTION:

The rotor is the rotating part of the motor's electromagnetic circuit. The most common type of rotor used in a three phase induction motor is a squirrel cage rotor. The squirrel cage rotor is so called because its construction is reminiscent of the rotating exercise wheel found in some pet cages. A squirrel cage rotor core is made by stacking thin steel laminations to form a cylinder. Rather than using coils of wire as conductors, conductor bars are die case into the slots evenly spaced around the cylinder. Most squirrel cage rotors are made by die casting aluminum to form the conductor bars. After die casting, rotor conductor bars are mechanically and electrically connected with end rings. The rotor is then pressed onto a steel shaft to form a rotor assembly.

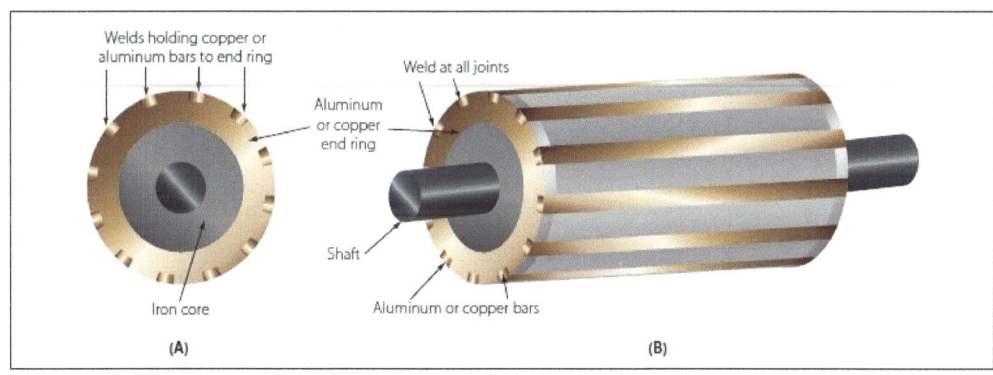

Squirrel cage rotor for an AC induction motor.

ENCLOSURE:

The enclosure consists of a frame and two end bells (or bearing housings) the stator is mounted inside the frame. The rotor fits inside the stator with a slight air gap separating it from the stator. There is no physical connection between the rotor and the stator. The enclosure protects the internal parts of the motor from water and other environmental elements. The degree of protection depends of the type of enclosure.

BEARINGS AND FAN:

Bearings mounted on the shaft, support the rotor and allow it to turn. Some motors use a fan also mounted on the rotor shaft to cool the motor when the shaft is rotating.

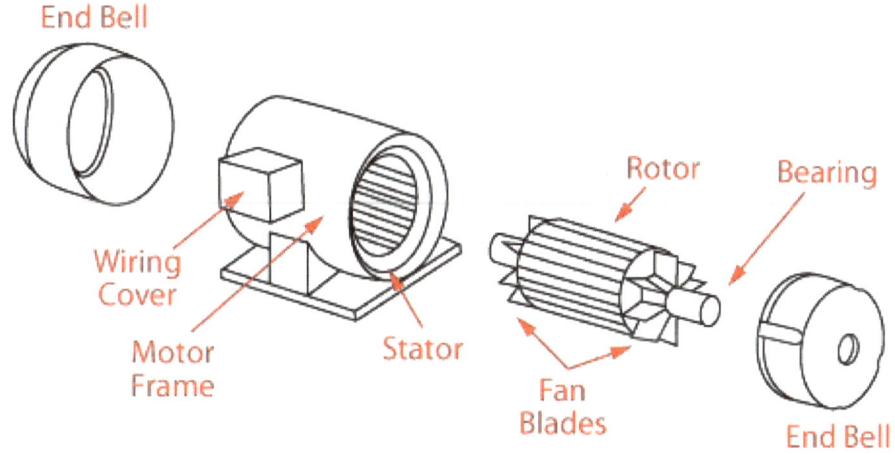

Induction Motor Advantages Disadvantages:

Almost 70% of the machines used in industries now a days are 3 phase induction motors. It works on the principle of induction where electro-magnetic field (emf) is induced in to the rotor conductors when rotating magnetic field of stator cuts the stationary rotor conductors. As the ac power is used in generation, transmission and distribution induction motors occupied significant place in industrial drive applications and out rule the dc motors which were earlier used for industrial applications. Induction motors are of two types based on the construction: Squirrel cage induction motor and slip ring induction motors. Squirrel cage induction motors are widely used in motor and drive applications. Some of the **advantages of induction motors** compared to dc motors and synchronous motors.
Also **disadvantages of induction motors** compared to other motors are also given below:

Induction Motor Advantages:

- Induction motors are simple and rugged in construction. Advantage of induction motors are that they are robust and can operate in any environmental condition
- Induction motors are cheaper in cost due to the absence of brushes, commutators, and slip rings
- They are maintenance free motors unlike dc motors and synchronous motors due to the absence of brushes, commutators and slip rings.
- Induction motors can be operated in polluted and explosive environments as they do not have brushes which can cause sparks

- 3 phase induction motors will have self starting torque unlike synchronous motors, hence no starting methods are employed unlike synchronous motor. However, single-phase induction motors does not have self starting torque, and are made to rotate using some auxiliaries.

These advantages in induction motors make them more prominent in industrial and domestic applications

Induction Motor Disadvantages:
Some of the disadvantages of induction motors compared to dc motors and synchronous motors are:

- 3 phase induction motors have poor starting torque and high have in rush currents. Therefore these motors are not widely used for applications which require high starting torques like traction systems. Squirrel cage induction motor has poor starting torque. Starting torque in the case of slip-ring induction motor is comparatively better because of the presence of external resistor in the rotor circuit during starting. Other important disadvantage of Induction motor is that it draws high inrush currents causing large momentary voltage dip during starting of the machine. High inrush currents can be reduced by employing some starting methods in induction motor
- Induction motors always operate under lagging power factor and during light load conditions they operate at very worst power factor (0.2 to 0.4 lagging). Some of the disadvantages of poor power are increase in I^2R losses in the system, reduction in the efficiency of the system. Hence some power factor correction equipments such as static capacitor banks should be placed near to these motors to deliver the reactive power to them.

- One of the main **disadvantages of induction motors** is that speed control of induction motors are difficult. Hence for fine speed control applications dc motors are used in place of induction motors. Due to advance in power electronics, variable frequency drives using induction motors are used in industries for speed control now a days.

These are some of the disadvantages associated with induction motors

Troubleshooting Induction Motor:

Use this resource to troubleshoot your AC motor. If the motor problems can't be resolved with this list, please contact *your supplier* for assistance.

1. MOTOR FAILS TO START UPON INITIAL INSTALLATION

- **Motor is wired incorrectly**
    - Refer to the wiring diagram to verify the motor is wired correctly.
- **Motor damaged and rotor is striking stator**
    - Rotate the motor's shaft and feel for rubbing.
- **Power supply or line trouble**
    - Check the source of power, overload, fuses, controls, etc..

2. MOTOR HAS BEEN RUNNING, THEN FAILS TO START

- **Fuse or circuit breaker is tripped**
    - Replace the fuse or reset the breaker.
- **Stator is shorted or went to ground (Motor will make a humming noise and the circuit breaker or fuse will trip)**
    - Check for leaks through the coils. If leaks are found, the motor must be replaced.
- **Motor overloaded or jammed**
    - Inspect to see that the load is free. Verify the amp draw of motor versus the nameplate rating.
- **Capacitor (on single phase motor) may have failed**

- First discharge the capacitor. To check the capacitor, set the volt-ohm meter to RX100 scale and touch its probes to the capacitor terminals. If the capacitor is OK, the needle will jump to zero ohms, and drift back to high. Steady zero ohms indicates a short circuit; steady high ohms indicates an open circuit.

3. MOTOR RUNS BUT DIES DOWN

- **Voltage drop**
  - If the voltage is less than 90% of the motor's rating, contact your power company or check to see that another piece of equipment isn't taking power away from the motor.
- **Load increased**
  - Verify that the load has not changed and the equipment has not gotten tighter. If it is a fan application, verify that the air flow hasn't changed.

4. MOTOR TAKES TOO LONG TO ACCELERATE

- **Defective capacitor**
  - Test the capacitor per previous instructions.
- **Bad bearings**
  - Noisy or rough feeling bearings should be replaced by the motor supplier.
- **Voltage too low**
  - Make sure the voltage is within 10% of the motor's nameplate rating. If not, contact your power company or check if some other equipment is taking power away from the motor.

5. MOTOR RUNS IN THE WRONG DIRECTION

- **Incorrect wiring**
  - Rewire the motor according to the schematic provided with the motor. Groschopp wiring diagrams can be found within the "Wiring Diagrams" page of our resources section or on individual motor pages.

6. MOTOR OVERLOADED/THERMAL PROTECTOR CONTINUOUSLY DRIPS

- **Load too high**
  - Verify that the load is not jammed. If the motor is a replacement, verify that the rating is the same as old motor. If the previous motor was a special design, a stock motor may not be able to duplicate the performance. Remove the load from the motor and inspect the amp draw of the motor unloaded. It should be less than the full load rating stamped on the nameplate (only true for three phase motors).
- **Ambient temperature too high**
  - Verify that the motor is getting enough air for proper cooling. Most motors are designed to run in an ambient temperature of or less than 40°C. (Note: A properly operating motor may be hot to the touch.)

7. MOTOR OVERHEATING

- **Overload. Compare actual amps (measured) with nameplate rating**
  - Locate and remove the source of excessive friction in the motor or load. Reduce the load or replace the motor with one of greater capacity.

- **Single phasing (three phase only)**
    - Check the current at all phases. It should be approximately the same.
- **Improper ventilation**
    - Check external cooling fan to be sure air is moving properly through the cooling channels. If there is excessive dirt build-up, clean the motor.
- **Unbalanced voltage (three phase only)**
    - Check the voltage at all phases. It should be approximately the same.
- **Rotor rubbing on stator**
    - Tighten the thru bolts.
- **Over voltage or under voltage**
    - Check the input voltage at each phase of the motor to make sure the motor is running at voltage specified on the nameplate.
- **Open stator winding (three phase only)**
    - Check the stator resistance at all three phases for balance.
- **Improper connections**
    - Inspect all the electrical connections for proper termination, clearance, mechanical strength, and electrical continuity. Refer to the motor lead diagram.

8. MOTOR VIBRATES

- **Motor misaligned to load**
    - Realign the load.
- **Load out of balance (direct drive application)**
    - Remove the motor from load and inspect the motor by itself. Verify that the motor shaft is not bent.
- **Defective motor bearings**

- o Test the motor by itself. If the bearings are bad, you will hear noises or feel roughness.
- **Load too light (single phase only)**
  - o Some vibration at a light load is standard. Consider switching to a smaller motor for excessive vibration.
- **Defective winding**
  - o Test the winding for shorted or open circuits. The amps may also be high. For defective winding, replace the motor.
- **High voltage**
  - o Check the power supply to make sure voltage is accurate.

9. BEARINGS FAIL

- **Load to motor may be excessive or unbalanced**
  - o Check the motor load and inspect the drive belt tension to ensure it's not too tight. An unbalanced load will also cause the bearings to fail.
- **High ambient temperatures**
  - o If the motor is used in an environment with high ambient temperatures, a different type of bearing grease may be required. You may need to consult the factory.
- **High motor temperatures**
  - o Check and compare the actual motor loads to the motor's rated load capabilities.

10. CAPACITOR FAIL

- **Ambient temperature too high**
  - o Verify that the ambient temperature does not exceed the motor's temperature rating (found on the nameplate)

- **Possible power surge to the motor (caused by a lightning strike or other high transient voltage)**
    - If this is a common problem, install a surge protector.

Basic Induction Motor Troubleshooting Chart:

Motor Problem	Cause	Remedy
Motor fails to start	Blown fuses	Replace fuse with proper type and rating
	Overload Trips	Check and reset overload in starter
	Improper power supply	Check to see that power supplied agrees with nameplate specifications and load factor
	Improper line connections	Check connections with wiring diagram supplied with motor
	Open circuit in winding or control switch	This is normally indicated by a humming sound when switch is closed. Check for loose wiring connections. Confirm that all control contacts are closing.
	Mechanical failure	Check to see that motor and drive turns freely. Check bearings and lubrication
	Short circuited stator	Indicated by blown fuses. Motor must be rewound
	Poor stator coil connections	Remove end belts. Locate poor connections with test lamp.
	Rotor defective	Check for broken bars or end rings
	Motor may be overloaded	Reduce motor load
Motor stalls	One phase may be open	Check supply lines for open phase

	Wrong application	Change type or size. Consult motor manufacturer
	Overload	Reduce load
	Low voltage	Check that nameplate voltage is maintained. Check connection.
	Open circuit	Fuses blown. Check overload relay, stator and push buttons
Motor runs and then dies down	Power failure	Check for loose connections to line, to fuses and to control
Motor does not come up to speed	Motor is applied for the wrong application	Consult manufacturer for right application of motor
	Voltage too low at motor terminals because of line drop	Use higher voltage on transformer terminals or reduce load. Check connections. Check conductors for proper size.
	Starting load too high	Check load motor is supposed to carry at start.
	Broken rotor bars or loose rotor	Look for cracks near the rings. A new rotor may be required as repairs are usually temporary not permanent
	Open primary circuit	Locate fault with testing device and repair.
Motor takes too long to accelerate and/or draws high	Excessive load	Reduce load
	Low voltage during start up	Check for high resistance. Adequate wire size.
	Defective squirrel cage rotor	Replace with new rotor

current (Amps)	Applied voltage too low	Improve voltage at terminals of transformer by tap changing.
Wrong rotation	Wrong sequence of phases	Reverse connections at motor or at switchboard.
Motor overheats while running under load	Overload	Reduce load
	Frame or bracket vents may be clogged with dirt and prevent proper ventilation of motor.	Open vent holes and check for a continuous stream of air from the motor.
	Motor may have one phase open	Check to make sure that all leads are well connected.
	Grounded coil	Locate and repair
	Unbalanced terminal voltage	Check for faulty leads, connections and transformers.
Motor vibrates	Motor misaligned	Realign
	Weak support	Strengthen base
	Coupling out of balance	Balance coupling
	Driven equipment unbalanced	Re-balance driven equipment
	Defective bearings	Replace bearing
	Bearings not in line	Line bearings up properly
	Balancing weights	Re-balance motor

	shifted	
	Poly-phase motor running single phase	Check for open circuit
	Excessive end play	Adjust bearing
Unbalanced line current on poly-phase motors during normal operation	Unequal terminal volts	Check leads and connections
	Single phase operation	Check for open contacts
	Unbalanced voltage	Correct unbalanced power supply
Noisy Operation	Airgap not uniform	Check and correct bracket fits or bearing.
	Rotor unbalance	Rebalance
Hot bearings general	Bent or sprung shaft	Straighten or replace shaft
	Excessive belt pull	Decrease belt tension
	Pulley too far away	Move pulley closer to motor bearing
	Pulley diameter too small	Use larger pulleys
	Misalignment	Correct by realignment of drive
Hot bearings ball	Insufficient grease	Maintain proper quantity of grease in bearing
	Deterioration of grease or lubricant contaminated	Remove old grease, wash bearings thoroughly in kerosene and replace with new grease.

	Excessive lubricant	Reduce quantity of grease, bearing should not be more than 1/2 filled
	Overloaded bearing	Check alignment, side and end thrust.
	Broken ball or rough races	Replace bearing, first clean housing thoroughly

Maintenance tips for Induction Motor:

The key to minimizing motor problems is scheduled routine inspection and service. The frequency of routine service varies widely between applications. Including the motors in the maintenance schedule for the driven machine or general plant equipment is usually sufficient. A motor may require additional or more frequent attention if a breakdown would cause health or safety problems, severe loss of production, damage to expensive equipment or other serious losses. Written records indicating date, items inspected, service performed and motor condition are important to an effective routine maintenance program. From such records, specific problems in each application can be identified and solved routinely to avoid breakdowns and production losses.

The routine inspection and servicing can generally be done without disconnecting or disassembling the motor. It involves the following factors:

Dirt and Corrosion

1. Wipe, brush, vacuum or blow accumulated dirt from the frame and air passages of the motor. Dirty motors run hot when thick dirt insulates the frame and clogged passages reduce cooling air flow. Heat reduces insulation life and eventually causes motor failure.
2. Feel for air being discharged from the cooling air ports. If the flow is weak or unsteady, internal air passages are probably clogged. Remove the motor from service and clean.
3. Check for signs of corrosion. Serious corrosion may indicate internal deterioration and/or a need for external repainting. Schedule the removal of the motor from service for complete inspection and possible rebuilding.

4. In wet or corrosive environments, open the conduit box and check for deteriorating insulation or corroded terminals. Repair as needed.

Lubrication

Lubricate the bearings only when scheduled or if they are noisy or running hot. Do NOT over-lubricate. Excessive grease and oil creates dirt and can damage bearings. See "Bearing Lubrication" for more details.

Heat, Noise and Vibration

Feel the motor frame and bearings for excessive heat or vibration. Listen for abnormal noise. All indicate a possible system failure. Promptly identify and eliminate the source of the heat, noise or vibration. See "Heat, Noise and Vibration" for details.

Winding Insulation

When records indicate a tendency toward periodic winding failures in the application, check the condition of the insulation with an insulation resistance test. See "Testing Windings" for details. Such testing is especially important for motors operated in wet or corrosive atmospheres or in high ambient temperatures.

BEARING LUBRICATION

Introduction

Modern motor designs usually provide a generous supply of lubricant in tight bearing housings. Lubrication on a scheduled basis, in conformance with the manufacturer's recommendations, provides optimum bearing life.

Thoroughly clean the lubrication equipment and fittings before lubricating. Dirt introduced into the bearings during lubrication probably causes more bearing failures than the lack of lubrication.

Too much grease can over pack bearings and cause them to run hot, shortening their life. Excessive
lubricant can find its way inside the motor where it collects dirt and causes insulation deterioration.

Many small motors are built with permanently lubricated bearings. They cannot and should not be lubricated.

Oiling Sleeve Bearings

As a general rule, fractional horsepower motors with a wick lubrication system should be oiled every 2000 hours of operation or at least annually. Dirty, wet or corrosive locations or heavy loading may require oiling at three-month intervals or more often. Roughly 30 drops of oil for a 3-inch diameter frame to 100 drops for a 9-inch diameter frame is sufficient. Use a 150 SUS viscosity turbine oil or SAE 10 automotive oil.

Some larger motors are equipped with oil reservoirs and usually a sight gage to check proper level.

As long as the oil is clean and light in color, the only requirement is to fill the cavity to the proper level with the oil recommended by the manufacturer. Do not overfill the cavity. If the oil is discolored, dirty or contains water, remove the drain plug. Flush the bearing with fresh oil until it comes out clean. Coat the plug threads with a sealing compound, replace the plug and fill the cavity to the proper level. When motors are disassembled, wash the housing with a solvent. Discard used felt packing. Replace badly worn bearings. Coat the shaft and bearing surfaces with oil and reassemble.

Greasing Ball and Roller Bearings

Practically all Reliance ball bearing motors in current production are equipped with the exclusive PLS/Positive Lubrication System. PLS is a patented open-bearing system that provides long, reliable bearing and motor life regardless of mounting position. Its special internal passages uniformly distribute new grease pumped into the housing during regreasing through the open bearings and forces old grease out through the drain hole. The close running tolerance between shaft and inner bearing cap minimizes entry of contaminants into the housing and grease migration into the motor. The unique V-groove outer slinger seals the opening between the shaft and end bracket while the motor is running or is at rest yet allows relief of grease along the shaft if the drain hole is plugged. (Figure 4)

The frequency of routine greasing increases with motor size and severity of the application as indicated in Table 1. Actual schedules must be selected by the user for the specific conditions.

During scheduled greasing, remove both the inlet and drain plugs. Pump grease into the housing using a standard grease gun and light pressure until clean grease comes out of the drain hole.

If the bearings are hot or noisy even after correction of bearing overloads (see "Troubleshooting") remove the motor from service. Wash the housing and bearings with a good solvent. Replace bearings that show signs of damage or wear. Repack the bearings, assemble the motor and fill the grease cavity.

Whenever motors are disassembled for service, check the bearing housing. Wipe out any old grease. If there are any signs of grease contamination or breakdown, clean and repack the bearing system as described in the preceding paragraph.

HEAT, NOISE AND VIBRATION

Heat

Excessive heat is both a cause of motor failure and a sign of other motor problems. The primary damage caused by excess heat is to increase the aging rate of the insulation. Heat beyond the insulation's rating shortens winding life. After overheating, a motor may run satisfactorily but its useful life will be shorter. For maximum motor life, the cause of overheating should be identified and eliminated. As indicated in the Troubleshooting Sections, overheating results from a variety of different motor problems. They can be grouped as follows:

1. WRONG MOTOR: It may be too small or have the wrong starting torque characteristics for the load. This may be the result of poor initial selection or changes in the load requirements.
2. POOR COOLING: Accumulated dirt or poor motor location may prevent the free flow of cooling air around the motor. In other cases, the motor may draw heated air from another source. Internal dirt or damage can prevent proper air flow through all sections of the motor. Dirt on the frame may prevent transfer of internal heat to the cooler ambient air.
3. OVERLOADED DRIVEN MACHINE: Excess loads or jams in the driven machine force the motor to supply higher torque, draw more current and overheat.

1. Light Duty: Motors operate infrequently (1 hour/day or less) as in portable floor sanders, valves, door openers.
2. Standard Duty: Motors operate in normal applications (1 or 2 work shifts). Examples include air conditioning units, conveyors, refrigeration apparatus, laundry machinery, woodworking and textile machines, water pumps, machine tools, garage compressors.
3. Heavy Duty: Motors subjected to above normal operation and vibration (running 24 hours/day, 365 days/year). Such operations as in steel mill service, coal and mining machinery, motor-generator sets, fans, pumps.
4. Severe Duty: Extremely harsh, dirty motor applications. Severe vibration and high ambient conditions often exist.
4. EXCESSIVE FRICTION: Misalignment, poor bearings and other problems in the driven machine, power transmission system or motor increase the torque required to drive the loads, raising motor operating temperature.
5. ELECTRICAL OVERLOADS: An electrical failure of a winding or connection in the motor can cause other Windings or the entire motor to overheat.

Noise and Vibration

Noise indicates motor problems but ordinarily does not cause damage. Noise, however, is usually accompanied by vibration.

Vibration can cause damage in several ways. It tends to shake windings loose and mechanically damages insulation by cracking, flaking or abrading the material. Embitterment of lead wires from excessive movement and brush sparking at commutators or current collector rings also results from vibration. Finally, vibration can speed bearing failure by causing balls to "brinnell," sleeve bearings to be pounded out of shape or the housings to loosen in the shells.

Whenever noise or vibration is found in an operating motor, the source should be quickly isolated and corrected. What seems to be an obvious source of the noise or vibration may be a symptom of a hidden problem. Therefore, a thorough investigation is often required.

Noise and vibrations can be caused by a misaligned motor shaft or can be transmitted to the motor from the driven machine or power transmission system. They can also be the result of either electrical or mechanical unbalance in the motor.

After checking the motor shaft alignment, disconnect the motor from the driven load. If the motor then operates smoothly, look for the source of noise or vibration in the driven equipment.

If the disconnected motor still vibrates, remove power from the motor. If the vibration stops, look for an electrical unbalance. If it continues as the motor coasts without power, look for a mechanical unbalance.

Electrical unbalance occurs when the magnetic attraction between stator and rotor is uneven around the periphery of the motor. This causes the shaft to deflect as it rotates creating a mechanical unbalance. Electrical unbalance usually indicates an electrical failure such as an open stator or rotor winding, an open bar or ring in squirrel cage motors or shorted field coils in synchronous motors. An uneven air gap, usually from badly worn sleeve bearings, also produces electrical unbalance. The chief causes of mechanical unbalance include a distorted mounting, bent shaft, poorly balanced rotor, loose parts on the rotor or bad bearings. Noise can also come from the fan hitting the frame, shroud, or foreign objects inside the shroud. If the bearings are bad, as indicated by excessive bearing noise, determine why the bearings failed. (See Troubleshooting Problems D and L.)

Brush chatter is a motor noise that can be caused by vibration or other problems unrelated to vibration. See Troubleshooting Problem M for details.

WINDINGS:

Care of Windings and Insulation

Except for expensive, high horsepower motors, routine inspections generally do not involve opening the motor to inspect the windings. Therefore, long motor life requires selection of the proper enclosure to protect the windings from excessive dirt, abrasives, moisture, oil and chemicals.

When the need is indicated by severe operating conditions or a history of winding failures, routine testing can identify deteriorating insulation. Such motors can be removed from service and repaired before unexpected failures stop production. See "Testing Windings".

Whenever a motor is opened for repair, service the windings as follows:

1. Accumulated dirt prevents proper cooling and may absorb moisture and other contaminants that damage the insulation. Vacuum the dirt from the windings and internal air passages. Do not use high pressure air because this can damage windings by driving the dirt into the insulation.

2. Abrasive dust drawn through the motor can abrade coil noses, removing insulation. If such abrasion is found, the winding should be revarnished or replaced.

3. Moisture reduces the dielectric strength of insulation which results in shorts. If

the inside of the motor is damp, dry the motor per information in "Cleaning and Drying Windings".

4. Wipe any oil and grease from inside the motor. Use care with solvents that can attack the insulation.

5. If the insulation appears brittle, overheated or cracked, the motor should be revarnished or, with severe conditions, rewound.

6. Loose coils and leads can move with changing magnetic fields or vibration, causing the insulation to wear, crack or fray. Revarnishing and retying leads may correct minor problems. If the loose coil situation is severe, the motor must be rewound.

7. Check the lead-to-coil connections for signs of overheating or corrosion. These connections are often exposed on large motors but taped on small motors. Repair as needed.

8. Check wound rotor windings as described for stator windings. Because rotor windings must withstand centrifugal forces, tightness is even more important. In addition, check for loose pole pieces or other loose parts that create unbalance problems.

9. The cast rotor rods and end rings of squirrel cage motors rarely need attention. However, open or broken rods create electrical unbalance that increases with the number of rods broken. An open end ring causes severe vibration and noise.

Testing Windings

Routine field testing of windings can identify deteriorating insulation permitting scheduled repair or replacement of the motor before its failure disrupts operations. Such testing is good practice especially for applications with severe operating conditions or a history of winding failures and for expensive, high horsepower motors and locations where failures can cause health and safety problems or high economic loss.

The easiest field test that prevents the most failures is the ground-insulation, or "megger," test. It applies DC voltage, usually 500 or 1000 volts, to the motor and measures the resistance of the insulation.

NEMA standards require a minimum resistance to ground at 40 degrees C ambient of 1 mega ohm per kv of rating plus 1 megohm. Medium size motors in good condition will generally have megohmmeter readings in excess of 50 mega ohms. Low readings may indicate a seriously reduced insulation condition caused by contamination from moisture, oil or conductive dirt or deterioration from age or excessive heat.

One megger reading for a motor means little. A curve recording resistance, with the motor cold and hot, and date indicates the rate of deterioration. This curve provides the information needed to decide if the motor can be safely left in service until the next scheduled inspection time.

The megger test indicates ground insulation condition. It does not, however, measure turn-to-turn insulation condition and may not pick up localized weaknesses. Moreover, operating voltage peaks may stress the insulation more severely than megger voltage. For example, the DC output of a 500-volt megger is below the normal 625-volt peak each half cycle of an AC motor operating on a 440-volt system. Experience and conditions may indicate the need for additional routine testing.

A test used to prove existence of a safety margin above operating voltage is the AC high potential ground test. It applies a high AC voltage (typically, 65% of a voltage times twice the operating voltage plus 1000 volts) between windings and frame.

Although this test does detect poor insulation condition, the high voltage can arc to ground, burning insulation and frame, and can also actually cause failure during the test. It should never be applied to a motor with a low megger reading.

DC rather than AC high potential tests are becoming popular because the test equipment is smaller and the low test current is less dangerous to people and does not create damage of its own.

Cleaning and Drying Windings

Motors which have been flooded or which have low megger readings because of contamination by moisture, oil or conductive dust should be thoroughly cleaned and dried. The methods depend upon available equipment.

A hot water hose and detergents are commonly used to remove dirt, oil, dust or salt concentrations from rotors, stators and connection boxes. After cleaning, the windings must be dried, commonly in a forced-draft oven. Time to obtain acceptable megger readings varies from a couple hours to a few days.

Precautions:

Read all of these basic safety precautions before starting to work on any of the motors. Failure to do so may result in personal injury, fire, and electric shock.

- You should always keep your work area clean and well lit.
- All motors include many small parts; keep very young children away from your work area.
- Motors spin very fast; always wear safety goggles.
- Super Glue, provided with most of the kits, is "instant bonding". Point tube away from face and body. Do not squeeze while opening.
- The T-pin and pushpin are sharp, be careful not to poke yourself.
- Do not leave the motor unattended.
- If you leave the battery shorted for a long period of time it may rupture or even explode.
- Do not exceed 12 Volts to power the motors. You may get an electric shock. High voltages may also cause parts to overheat and create a fire.
- If the current flowing through the transistor is big, it may get very hot. Do not touch it to prevent burns.
- The soldering iron gets very hot! Be very careful with this tool, as it is easy to burn yourself with it.
- Health hazard warning: Solder may contain lead. It is not suitable for small children.

These motor kits should not be used by anyone younger than 13 years old without adult supervision.

End of Book

www.ingramcontent.com/pod-product-compliance
Lightning Source LLC
Chambersburg PA
CBHW042323250526
R18347300001B/R183473PG45473CBX00021B/17